THE MANIFESTATION PLAYBOOK

Olivia Praise

Table of Contents

INTRODUCTION 5
CHAPTER ONE 7
Getting your Raspberry Pi up and running 7
    Start or boot Media 9
    Keyboard 11
    Mouse 12
    Display 13
    Audio 14
CHAPTER TWO 16
Set up the operating system 16
    Install with imager 18
    OS personalization 23
    Configure throughout the network 34
CHAPTER THREE 39
Configure the Raspberry Pi. 39
    Initial boot configuration 41
    Bluetooth 43
    Place 43
    User 44
    Wi-Fi 45
    Browser 47
    Software updates 48
    Finish 50

Copyright 2023 © Olivia Praise

All rights reserved. This book is copyright and no part of it may be reproduced, stored or transmitted, in any form or means, without the prior written permission of the copyright owner.

Printed in the United States of America

Copyright 2023 © Olivia Praise

INTRODUCTION

The Raspberry Pi Foundation unveiled the fifth edition of the device, which includes several significant improvements.

The Raspberry Pi 4 was released more than four years ago, and for a while, there was a sense that the wait for the Raspberry Pi 5 may go on forever.

But for those looking for a more powerful Pi, the Raspberry Pi foundation has been released with some significant hardware enhancements.

The new I/O chip, which was created internally by the Raspberry Pi Foundation, and the entirely new CPU are the two biggest upgrades.

In addition, the new Raspberry Pi 5 has a far higher power draw than previous models. It is more effective at the same workloads, but it may drive the cores much harder.

This results in two more issues. Initially, you'll probably require better cooling, which might be achieved with an active cooling fan or a large passive heat sink. Second, the Pi 4 power supply will not pump enough power to get the best performance out of the new Pi; you will need a new one.

An RTC chip will also be integrated into the Pi 5. However, in order for this to work in the event that your Pi loses power, an additional battery has to be connected in. The only upside is that you won't have to purchase an RTC chip separately.

CHAPTER ONE

Getting your Raspberry Pi up and running

In order to begin using your Raspberry Pi, you will require the following:

• A power source

• Boot media, such as a fast and large-capacity microSD card

Your Raspberry Pi may be configured as a headless computer that can only be accessed over a network, or as an interactive computer with a desktop. You don't need any more accessories to set up your Raspberry Pi headless; during operating system installation, you may preconfigure a domain name, user account, network connection, and SSH. To use your Raspberry Pi straight, you will want the following extra parts:

- A display

- A wire to link your monitor and Raspberry Pi
- A keyboard
- A mouse

Power Source or supply

The USB-PD power mode needed to power different Raspberry Pi models is displayed in the following table. Any excellent power supply that offers the appropriate power mode is suitable for usage.

Model	Recommended Power Supply (Voltage/Current)	Raspberry Pi Power Supply
Raspberry Pi 5	5V/5A, 5V/3A limits peripherals to 600mA	27W USB-C Power Supply
Raspberry Pi 4 Model B	5V/3A	15W USB-C Power Supply
Raspberry Pi 3 (all models)	5V/2.5A	12.5W Micro USB Power Supply
Raspberry Pi 2 (all models)	5V/2.5A	12.5W Micro USB Power Supply
Raspberry Pi 1 (all models)	5V/2.5A	12.5W Micro USB Power Supply
Raspberry Pi Zero (all models)	5V/2.5A	12.5W Micro USB Power Supply

Connect the connector labeled "POWER IN," "PWR IN," or "PWR" to your power supply. Certain Raspberry Pi models, like the Zero series, contain USB connections for output that are identical in shape to the power port. Make sure your Raspberry Pi is plugged into the right port!

Start or boot Media

Models of Raspberry Pi that don't have onboard storage need you to supply it. Any compatible media, including USB drives, network drives, PCIe HAT-connected storage, and microSD cards, may be used to boot your Raspberry Pi from an operating system image. All of these media kinds are supported by newer Raspberry Pi models only, though.

Since the Raspberry Pi 1 variant A+, every consumer variant of the Raspberry Pi has included a microSD slot. When a card is inserted into the microSD slot, your Raspberry Pi starts up immediately from that location.

When running Raspberry Pi OS on an SD card, we advise using one that has at least 16GB of capacity.

For Raspberry Pi OS Lite users, we advise using at least 4GB.

The following devices may only be booted from a boot partition of 256GB or less due to a hardware limitation:

• Early Raspberry Pi 2 variants with the BCM2836 SoC; Raspberry Pi Zero; Raspberry Pi 1;

The prerequisites vary for various operating systems. For information on capacity requirements, consult your operating system's manual.

Keyboard

You may attach a wired keyboard or a USB Bluetooth receiver to any one of your Raspberry Pi's USB ports.

Mouse

You may add a corded mouse or USB Bluetooth receiver to your Raspberry Pi using either of its USB ports.

Display

Connect your primary display to the HDMI0 port on your Raspberry Pi if it has multiple HDMI ports.

Some Raspberry Pi devices may connect to displays using the following:

Model	Display outputs
Raspberry Pi 5	2x micro HDMI
Raspberry Pi 4 (all models)	2x micro HDMI, audio and composite out via 3.5mm TRRS jack
Raspberry Pi 3 (all models)	HDMI, audio and composite out via 3.5mm TRRS jack
Raspberry Pi 2 (all models)	HDMI, audio and composite out via 3.5mm TRRS jack
Raspberry Pi 1 Model B+	HDMI, audio and composite out via 3.5mm TRRS jack
Raspberry Pi 1 Model A+	HDMI, RCA connector
Raspberry Pi Zero (all models)	mini HDMI

The majority of screens lack micro or tiny or mini HDMI connectors. To connect those ports on your Raspberry Pi to any HDMI monitor, you may use a micro HDMI or mini HDMI to HDMI cable. If your display does not support HDMI, you might think about getting an adapter that converts HDMI display output to a port that your display can handle.

Audio

Audio output over HDMI is supported by all Raspberry Pi models equipped with HDMI, micro HDMI, or mini HDMI. Every Raspberry Pi model has USB audio capability. All Raspberry Pi devices with Bluetooth are capable of playing music via Bluetooth. The 3.5mm auxiliary TRRS connection on all Raspberry Pi 1, 2, 3, and 4 models may need to be amplified in order to produce a loud enough output.

Linked In

The Raspberry Pi versions listed below include Bluetooth and WiFi connectivity:

- Raspberry Pi 5
- Raspberry Pi 4
- Raspberry Pi 3B+
- Raspberry Pi 3
- Raspberry Pi Zero W
- Raspberry Pi Zero 2 W

A variation having an Ethernet port is indicated by the suffix "Model B"; those without one are indicated by "Model A". A USB-to-Ethernet converter can be used to establish a wired internet connection even if your model lacks an Ethernet connector.

CHAPTER TWO

Set up the operating system.

An operating system is necessary for you to operate your Raspberry Pi. Any SD card that is placed into the SD card slot causes Raspberry Pis to automatically search for an operating system on it.

You may boot an operating system from various storage devices, such as USB drives, storage linked via a HAT, and network storage, depending on the type of Raspberry Pi you're using.

For the purpose of installing an operating system on a hard drive for your Raspberry Pi, you will require the following: a computer that you can utilize to image the storage disk into a boot device; and a means of connecting the storage device to that computer.

MicroSD cards are the boot device of choice for most Raspberry Pi users.

We advise using Raspberry Pi Imager to install an operating system.

With Linux, Windows, and macOS, you may download and write pictures with the aid of a program called Raspberry Pi Imager. Several well-known Raspberry Pi operating system images are included in Imager. Additionally, Imager may load images that have been obtained straight from the Raspberry Pi or from other sources like Ubuntu. To preconfigure your Raspberry Pi's credentials and remote access settings, utilize Imager.

Imager is compatible with container formats such as.zip and supports images encoded in the.img format.

Should you be without a computer to copy an image on a boot drive, you may be able to set up an operating system via the internet straight onto your Raspberry Pi.

Install with imager

Imager may be installed using the methods listed below:

• Run the installer after downloading the most recent version from raspberrypi.com/software.

• Install it using your package manager from a terminal (sudo apt install rpi-imager, for example).

After installing Imager, run rpi-imager or hit the Raspberry Pi Imager logo to open the program.

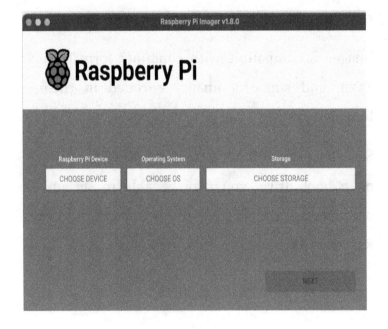

To pick your Raspberry Pi model via the available options, hit Choose device.

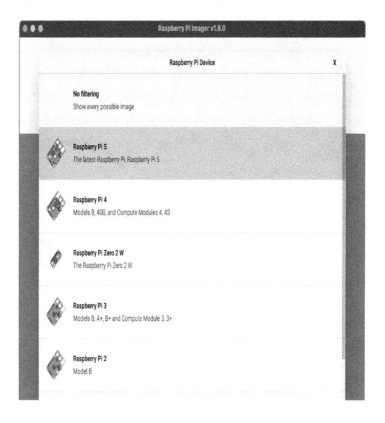

After that, hit Choose OS to pick an installation operating system. The Raspberry Pi OS version that Imager recommends for your model is always shown first in the list of options.

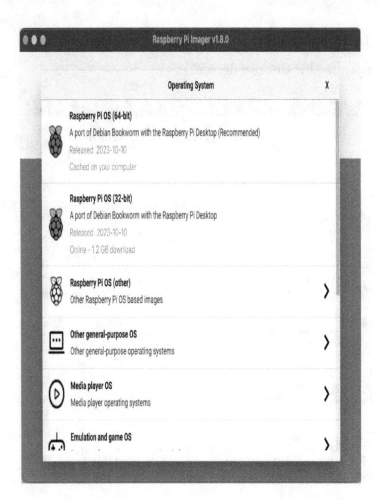

Link your computer to the storage device of your choice. For instance, use an external or integrated SD card reader to insert a microSD card. Next, choose your storage device by clicking Choose storage.

Make sure you select the right storage device if you have many devices connected to your computer! Storage devices may frequently be identified by size. Disconnect other devices until you've determined which device you want to image if you're not sure.

Hit Write after that.

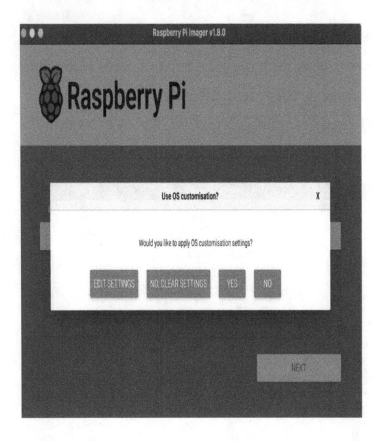

Imager will prompt you to set up OS customization in a popup. Setting up your Raspberry Pi using the OS customization options is highly advised. To access the OS customization, tap the Edit Settings icon.

Should you choose not to customize your Raspberry Pi using the OS customization options, Raspberry Pi

OS will prompt you for the same data during the configuration wizard's initial boot. If you would like to forego OS customization, choose the No box.

OS personalization

Prior to initial startup, you can configure your Raspberry Pi using the OS customization menu. Preconfigure is an option.

- A username and password
- Wi-Fi credentials
- The device hostname
- The time zone
- Your keyboard layout
- Remote connectivity

You may get a popup asking for authorization to load WiFi credentials via your host computer after you first access the OS customization menu. Imager will automatically fill in your WiFi credentials using the network you are presently connected to if you answer "yes". You can manually input your WiFi credentials if you answer "no."

The hostname setting establishes the hostname that your Raspberry Pi uses for mDNS broadcasts to the network. Using <hostname>.local or <hostname>.lan, other networked devices can connect to your Raspberry Pi and communicate with your computer.

On your Raspberry Pi, the admin user account's username and password are specified using the username and password option.

You may configure your wireless network by entering its password and SSID (name) using the wireless LAN option. You should activate the "Hidden SSID" feature if your network isn't sending out an SSID publicly. Imager sets the "Wireless LAN country" to where you are currently by default. The WiFi broadcast frequencies that your Raspberry Pi uses are managed by this parameter. If you intend to use a headless Raspberry Pi, provide the login credentials for the wireless LAN option.

You may specify your Pi's time zone and preferred keyboard layout using the locale settings option.

You may configure the Services tab to enable remote Raspberry Pi connections.

Select the Enable SSH option if you intend to use your Raspberry Pi remotely across your network. In case you intend to use a headless Raspberry Pi, you ought to activate this feature.

• Select the password authentication option and use your username and password from the general tab of OS customization to SSH access your Raspberry Pi over the network.

- Using a private key from the computer you are presently using, enable your Raspberry Pi to allow passwordless public-key SSH authentication by selecting Allow public-key authentication only. Imager uses the public key that you already have if you have an RSA key set up for SSH. If not, you may create a public/private key pair by clicking Run SSH-keygen. The freshly created public key will be used by Imager.

Additionally, an Options menu that lets you adjust Imager's behavior during a write is part of the OS customization. With these options, you may turn off telemetry, immediately unmount storage media after image verification, and play a noise once Imager finishes validating an image.

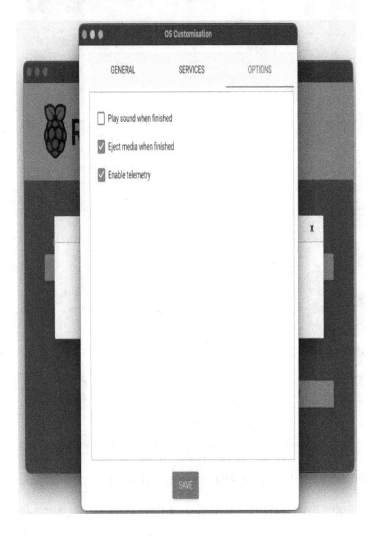

Hit Save, to enable you save your customization when you've completed inputting the OS customization options.

After that, as you write the image to your storage device, select "Yes" to apply the OS customization options.

To finally start writing data to the storage device, answer "Yes" when prompted by the "Are you pretty sure you would want to proceed or continue?" box.

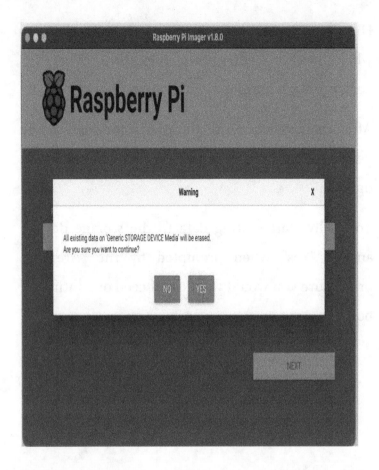

It's okay to continue if you encounter an admin prompt requesting authorization for both reading and writing your storage media.

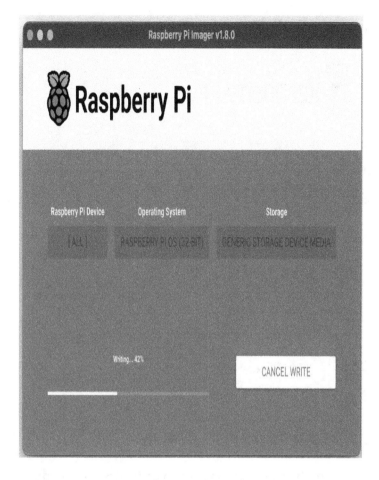

Take a stroll or get a cup of coffee. This can need several minutes.

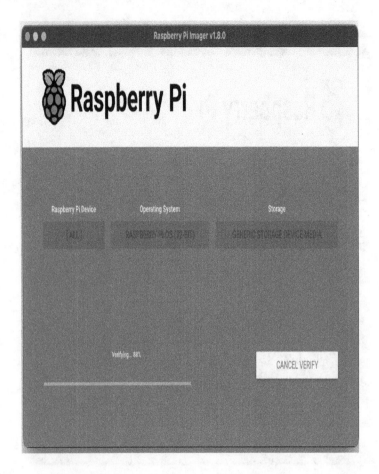

You can choose to bypass the verification procedure by clicking cancel verify if you so wish to live exceptionally riskily.

The "Write Successful" window indicates that the picture has been fully written and validated. Now that the storage device is ready, you may boot a Raspberry Pi from it!

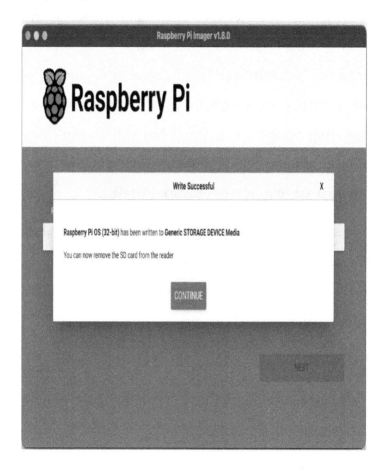

For the purpose of getting your Raspberry Pi operating, continue to the initial boot setup procedures.

Configure throughout the network

With a version of Raspberry Pi Imager downloaded over the network, Network Install allows a Raspberry Pi to set up an operating system on a storage device. Without an additional SD card reader or computer other than your Raspberry Pi, you may install an operating system on it using Network Install. Any suitable storage device, including USB and SD cards, can be used to execute Network Install.

Runs only on Raspberry Pi 4, 400 for Network Install. To utilize Network Install, you might need to upgrade the bootloader on your Raspberry Pi if it is running an outdated version.

The following are necessary for Network Install:

• A Raspberry Pi model that is suitable and running network install firmware

- A monitor
- A keyboard
- A wired internet connection

Turn on your Raspberry Pi and hold down the SHIFT key in the settings shown below to initiate Network Install:

• No removable boot disk

• Connected keyboard; • connected suitable storage device, like a USB drive or SD card

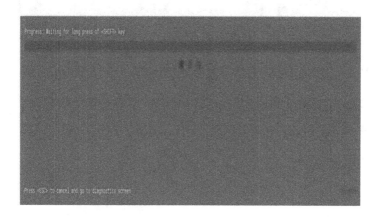

Use an Ethernet wire to connect your Raspberry Pi to the internet assuming you aren't already.

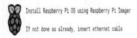

Your Raspberry Pi will start downloading the Raspberry Pi installer as soon as it is online. You can try again by repeating the procedure if the download doesn't work.

Your Raspberry Pi will launch Raspberry Pi Imager immediately when you have finished downloading Raspberry Pi Installer.

Raspberry Pi

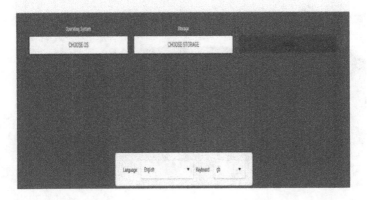

CHAPTER THREE

Configure the Raspberry Pi.

Once the operating system image has been installed, attach your storage device quickly to your Raspberry Pi.

To make sure the Raspberry Pi is switched off while you attach peripherals, first disconnect the power supply. You may now insert the microSD card that you used to install the operating system into the Raspberry Pi's card slot. You may now attach any additional storage device that you have installed the operating system on to your Raspberry Pi.

Next, connect any other accessories, including your keyboard and mouse.

Lastly, attach your Raspberry Pi's power supply to it. When your Pi starts up, the status LED ought to illuminate. You should see the startup screen on your Pi in a matter of minutes if it is connected to a monitor.

Initial boot configuration

Congratulations if you preconfigured your Raspberry Pi using OS modification in Imager! You may use your device now. To find out how to get the most of your Raspberry Pi, continue to the next stages.

Examine the status LED if your Raspberry Pi doesn't boot up in five minutes.

If your Pi won't start up, try these troubleshooting steps:

Try launching from an SD card if you previously used a different boot device. Re-image your SD card, making careful to finish Imager's verification process. Update your Raspberry Pi's bootloader before re-imaging your SD card.

Upon initial boot, a configuration wizard will be launched on your Raspberry Pi if you decided not to customize the OS using Imager. To utilize the wizard, you'll need a keyboard and monitor; a mouse is not necessary.

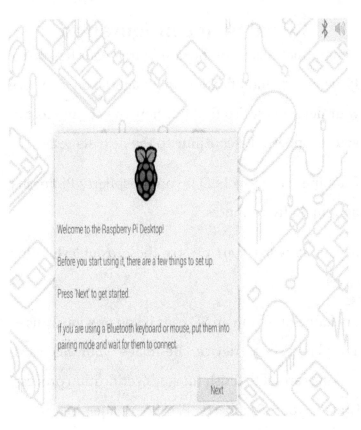

Bluetooth

This step will let you couple your smartphone with a Bluetooth keyboard or mouse. When it detects a device that can be paired, your Raspberry Pi will connect to the first one it discovers.

Both internal and external USB Bluetooth adapters are compatible with this method. Make sure to connect in any USB adapters before starting your Raspberry Pi.

Place

You may set your keyboard layout, language, nation, and time zone on this page.

User

You can configure the password and username for the user account that you are using.

Older versions of the OS for the Raspberry Pi default to having the username "pi". If you use the "pi" instead of the "raspberry" password, it will prevent unauthorized access to your device.

Wi-Fi

You can choose the network that's ideal for you from the list provided.

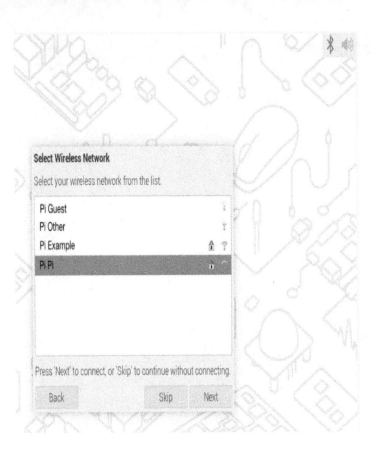

You can enter the password for your network if it requires one.

Browser

You can select either Chromium or Firefox as your default browser. You can also remove the browser that you don't want.

Software updates

After you have internet access, you can download the latest version of your Raspberry Pi's software and operating system. This process will automatically update your device's software and settings. You can then choose to remove the non-essential browser by going to the "browser selection" option and choosing

"Uninstall." It will take a few minutes to update your device's software.

After you see a message that says that your system is currently up-to-date, click OK and proceed.

Finish

After you've completed the configuration wizard, you'll be prompted to reboot the Raspberry Pi. It will then automatically apply the latest configuration and boot to your desktop.

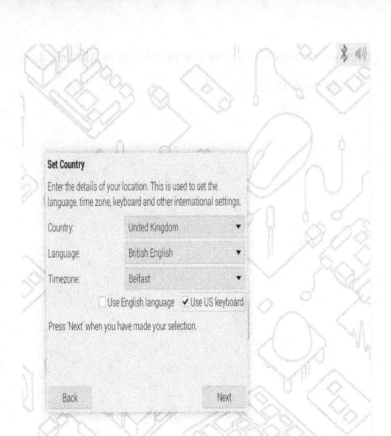

www.ingramcontent.com/pod-product-compliance
Lightning Source LLC
LaVergne TN
LVHW012317060125
800679LV00028B/1178